D1576264

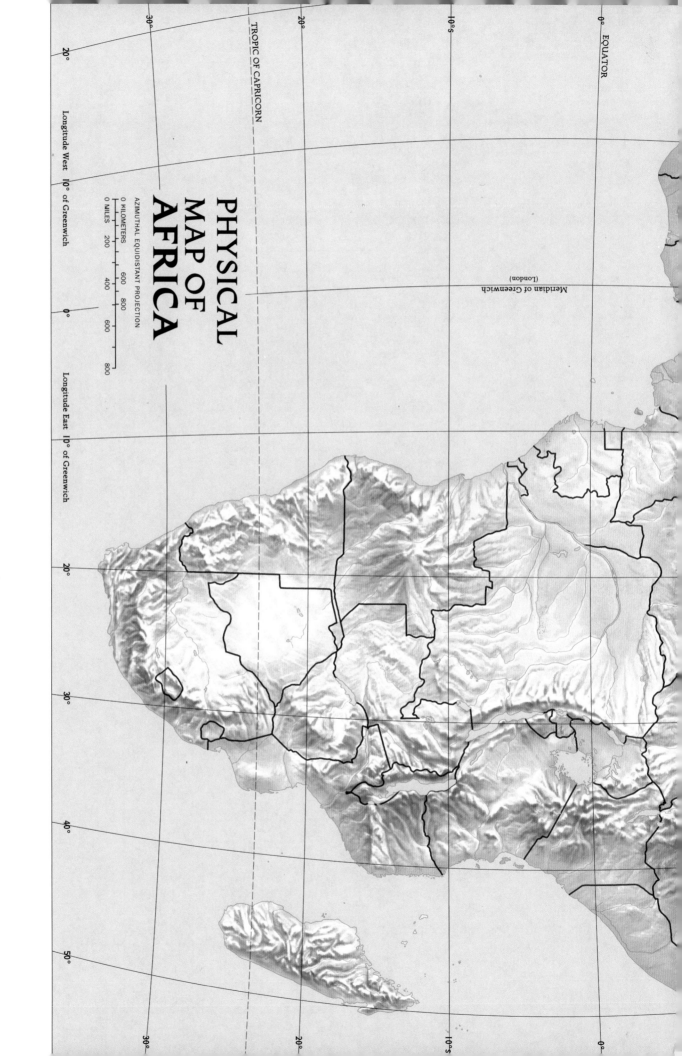

PHYSICAL MAP OF AFRICA

AZIMUTHAL EQUIDISTANT PROJECTION

0 KILOMETERS   200   400   600   800

0 MILES   200   400   600   800

EQUATOR

Meridian of Greenwich
(London)

TROPIC OF CAPRICORN

Longitude West   10°   of Greenwich   0°   Longitude East   10°   of Greenwich

0°

10°S

20°

10°S

20°

30°

30°

20°

10°S

0°

20°

30°

40°

50°

30°

20°

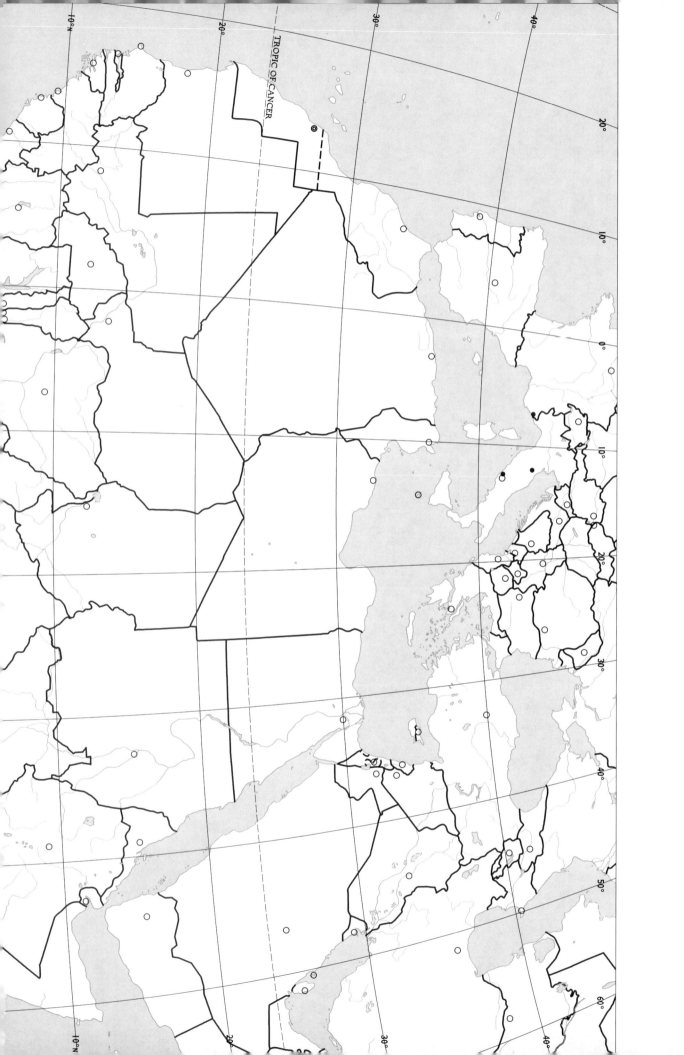

TROPIC OF CANCER

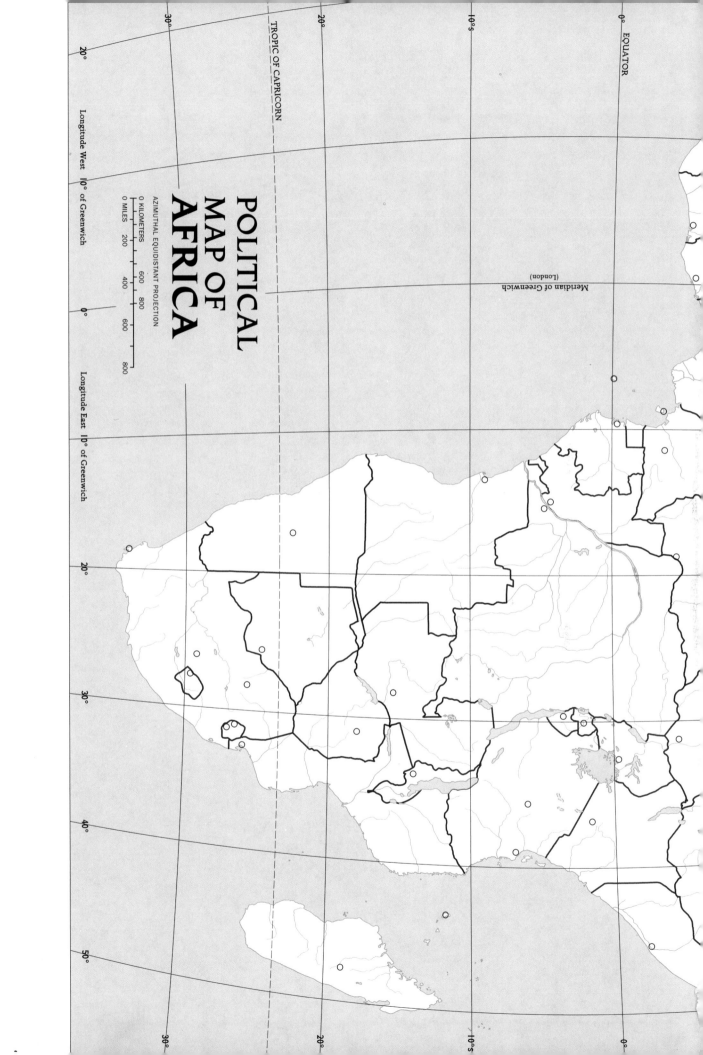

POLITICAL
MAP OF
AFRICA

AZIMUTHAL EQUIDISTANT PROJECTION

0 KILOMETERS
200   600   800   600   800
0 MILES   400   800

EQUATOR

Meridian of Greenwich
(London)

TROPIC OF CAPRICORN

Longitude West 10° of Greenwich

0°

Longitude East 10° of Greenwich

20°

10°S

0°

20°

30°

20°

30°

40°

50°

30°

20°

10°S

0°

150°   160°   170°   180°

Monday | Sunday

EQUATOR   0°

Date Line

10°S

20°

TROPIC OF CAPRICORN

30°

40°

150°   160°   170°   180°

110°   120°   130°   140°

0°

10°S

20°

30°

40°

PHYSICAL MAP OF
AUSTRALIA
AND Oceania

MERCATOR PROJECTION

0 KILOMETERS   500   1000
0 MILES   500   1000

110°   120°   130°   Longitude East 140° of Greenwich

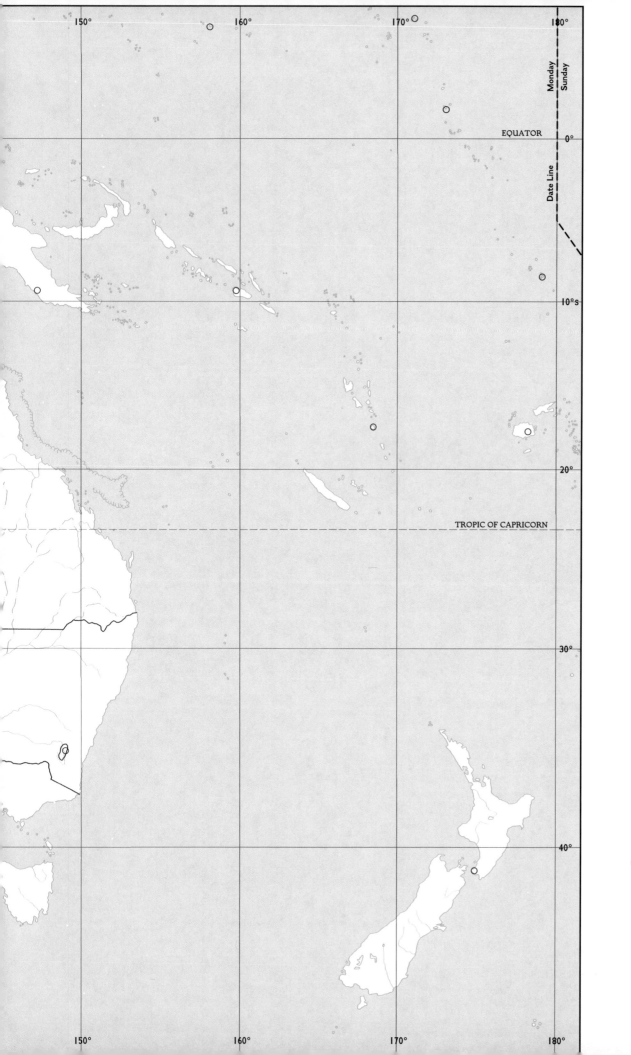

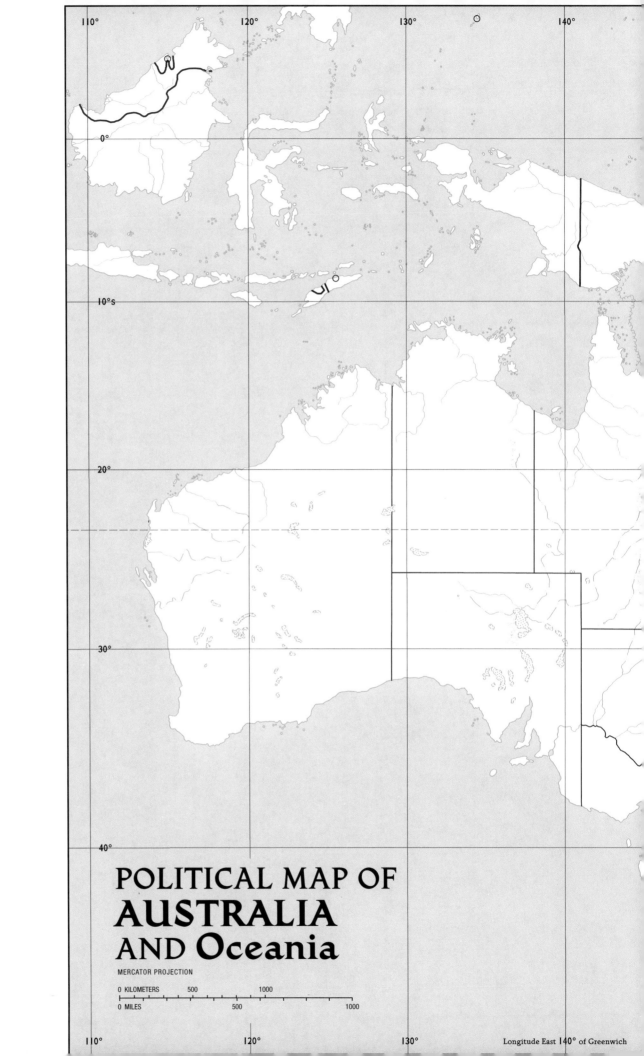

POLITICAL MAP OF
AUSTRALIA
AND Oceania

MERCATOR PROJECTION

0 KILOMETERS   500   1000

0 MILES   500   1000

110°   120°   130°   140°

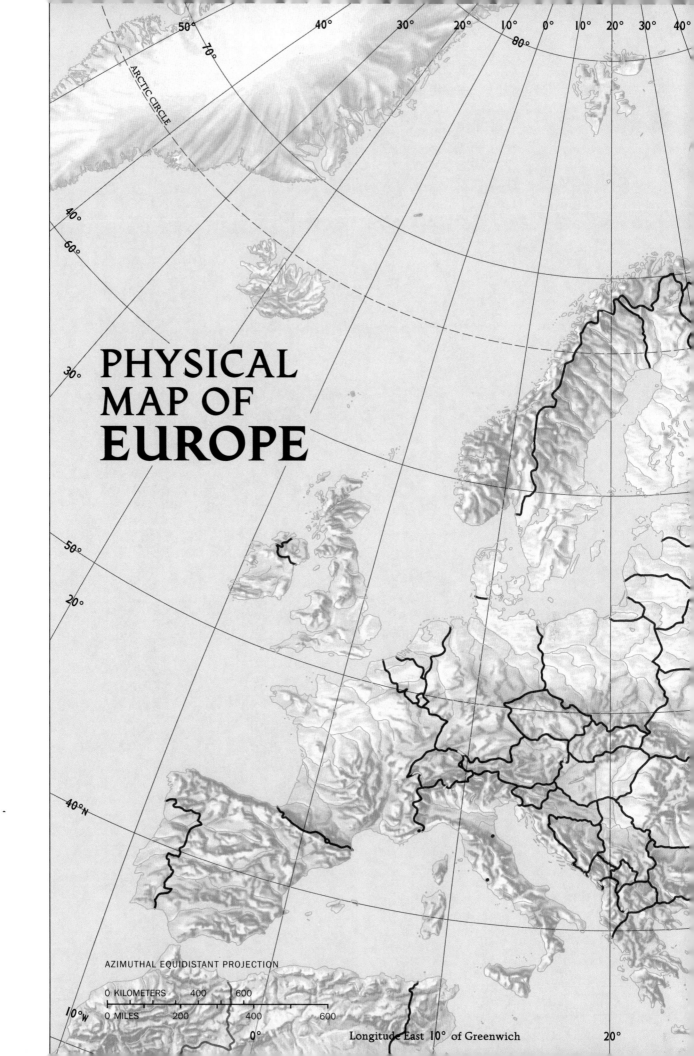

PHYSICAL
MAP OF
EUROPE

ARCTIC CIRCLE

AZIMUTHAL EQUIDISTANT PROJECTION

| 0 KILOMETERS | 400 | 600 |

| 0 MILES | 200 | 400 | 600 |

Longitude East 10° of Greenwich

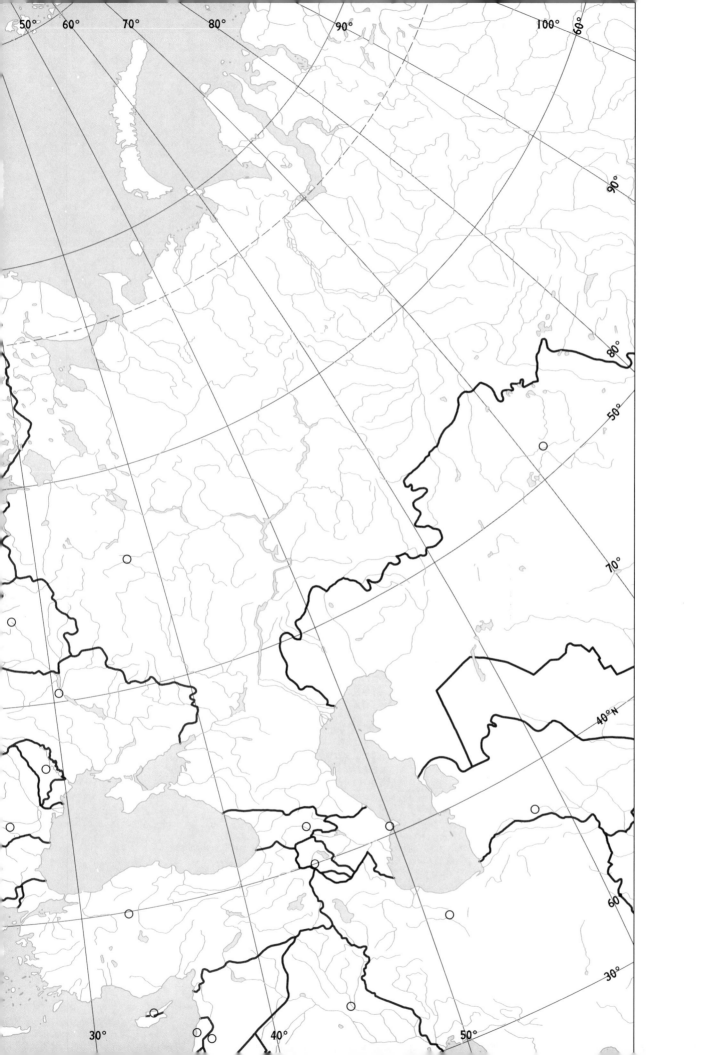

POLITICAL
MAP OF
EUROPE

ARCTIC CIRCLE

AZIMUTHAL EQUIDISTANT PROJECTION

| 0 KILOMETERS | 400 | 600 |
| 0 MILES | 200 | 400 | 600 |

Longitude East 10° of Greenwich

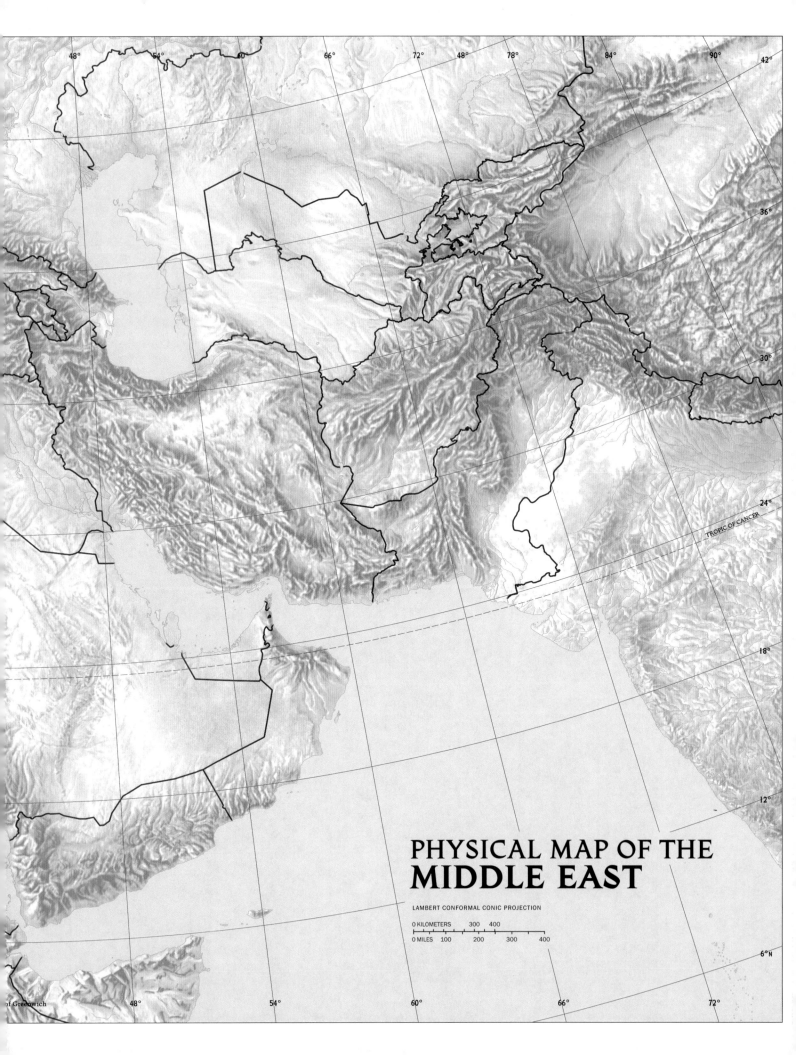

48° 54° 60° 66° 72° 48° 78° 84° 90° 42°

36°

30°

TROPIC OF CANCER 24°

18°

12°

PHYSICAL MAP OF THE
MIDDLE EAST

LAMBERT CONFORMAL CONIC PROJECTION

0 KILOMETERS   300   400

0 MILES   100   200   300   400

6°N

of Greenwich   48°   54°   60°   66°   72°

42°

6°   0°   6°   48°   12°   18°   24°   30°   36°   42°

36°

30°

24°

18°

12°

6° N

12°   18°   24°   30°   36°   Longitude East 42°

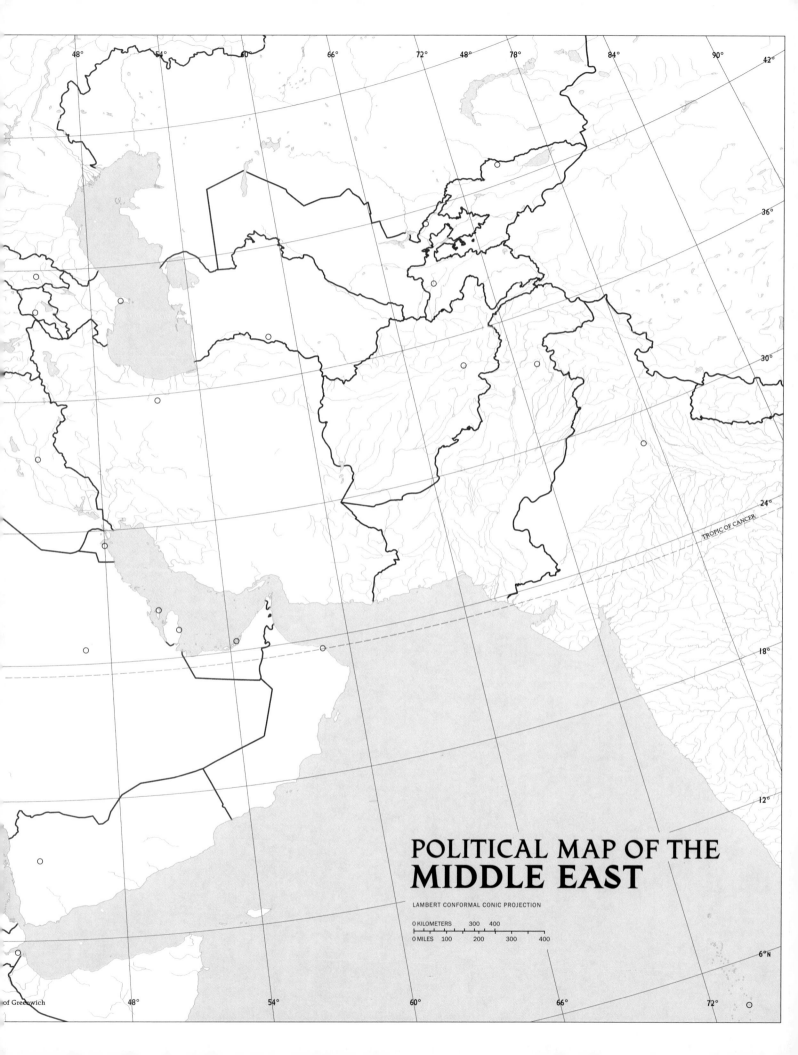

POLITICAL MAP OF THE
MIDDLE EAST

LAMBERT CONFORMAL CONIC PROJECTION

0 KILOMETERS   300 400
0 MILES   100   200   300   400

TROPIC OF CANCER

of Greenwich

42°
36°
30°
24°
18°
12°
6° N

48°   54°   60°   66°   72°   48°   78°   84°   90°
48°   54°   60°   66°   72°

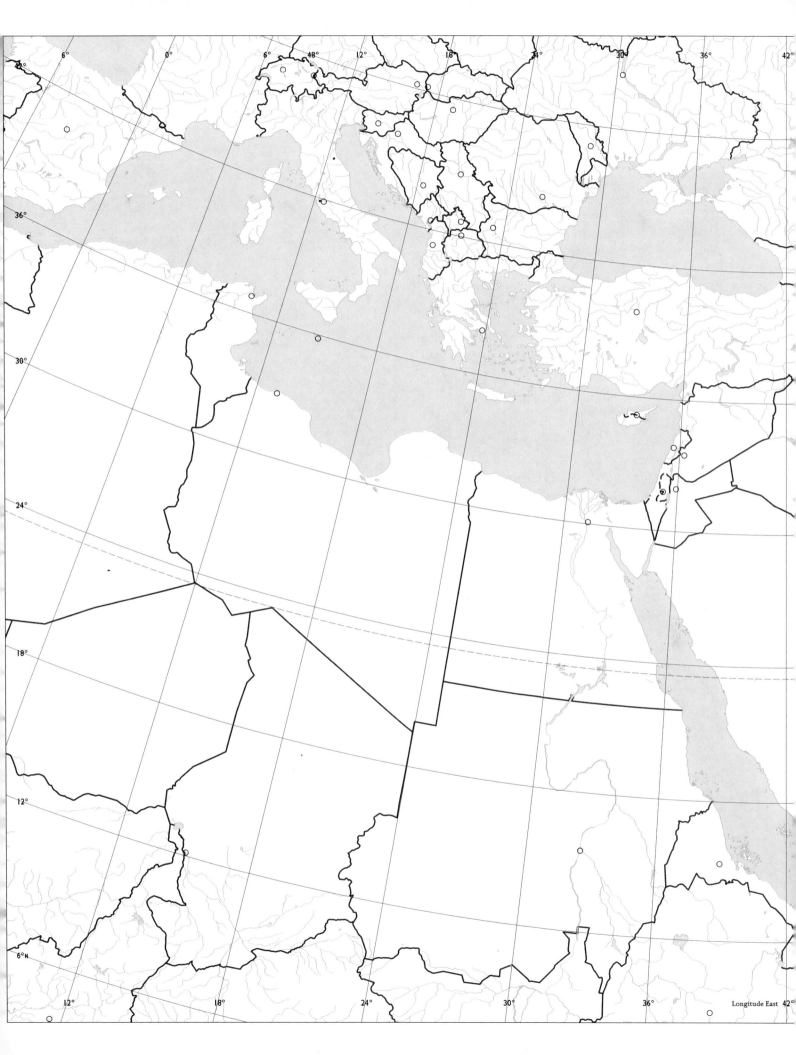

84°   78°   72°   66°   60°   54°   30°

24°

TROPIC OF CANCER

18°

12°

6°N

60°

EQUATOR   0°

84°   78°   72°   66°

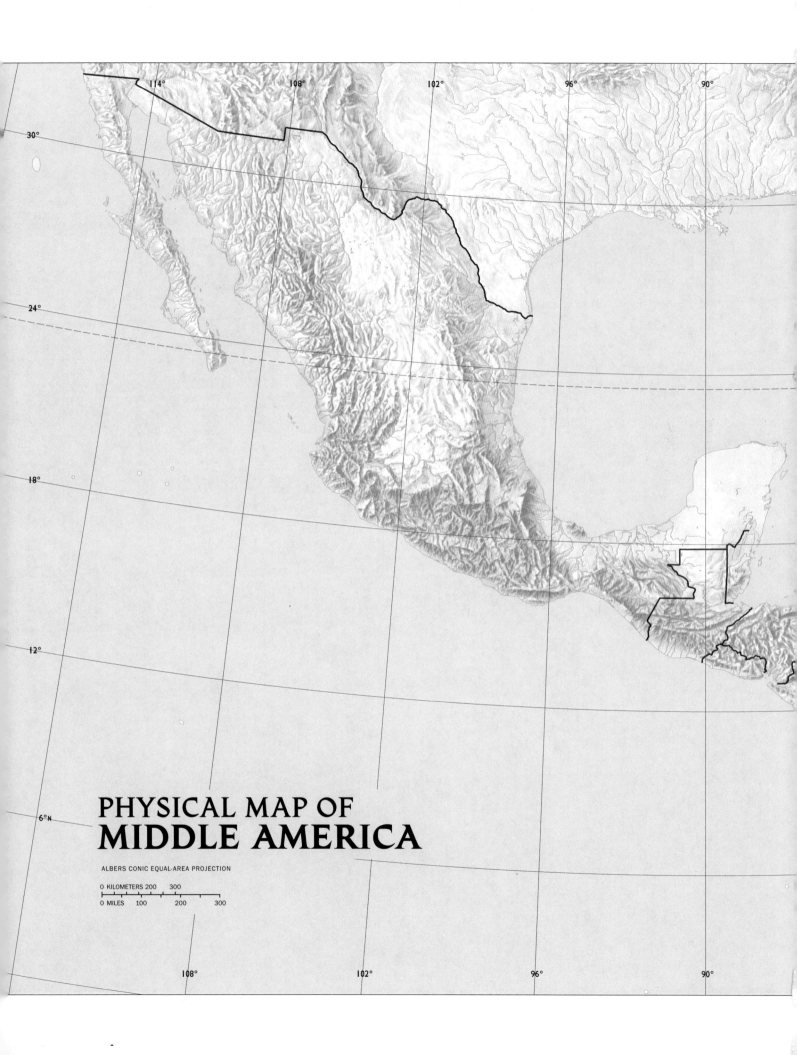

PHYSICAL MAP OF
MIDDLE AMERICA

ALBERS CONIC EQUAL-AREA PROJECTION

0 KILOMETERS 200   300

0 MILES   100   200   300

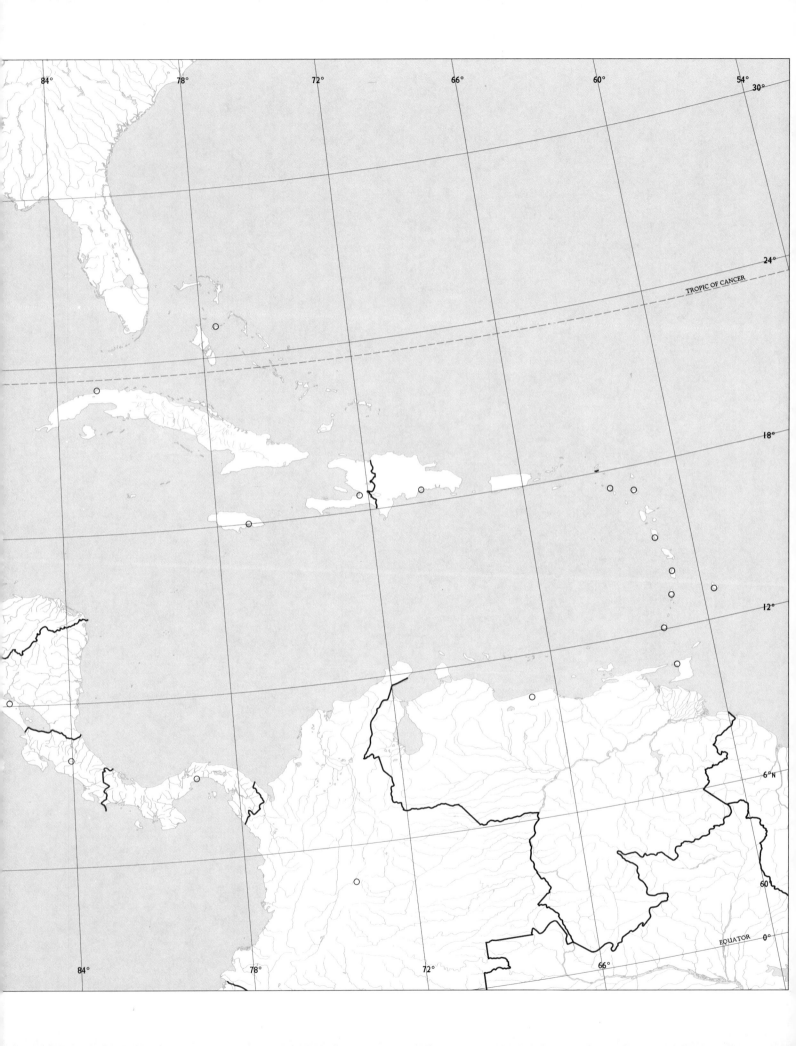

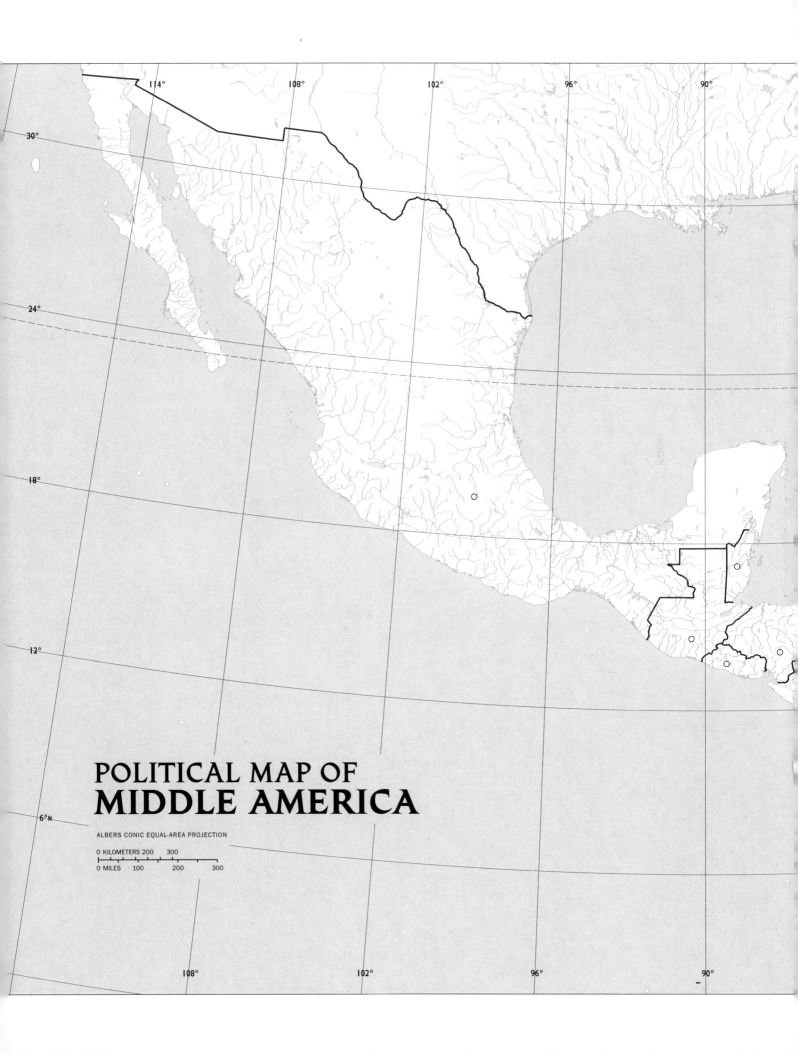

POLITICAL MAP OF
MIDDLE AMERICA

ALBERS CONIC EQUAL-AREA PROJECTION

0 KILOMETERS 200   300

0 MILES   100   200   300

ARCTIC CIRCLE

North Pole

Meridian of Greenwich
(London)

PHYSICAL MAP OF
NORTH AMERICA

AZIMUTHAL EQUIDISTANT PROJECTION

0 KILOMETERS   200   400   600   800   600   800

0 MILES   400   600

Longitude West 100° of Greenwich

TROPIC OF CANCER

130°

20°N

10°N

30°

110°

90°

80°

70°

20°N

30°

60°

40°

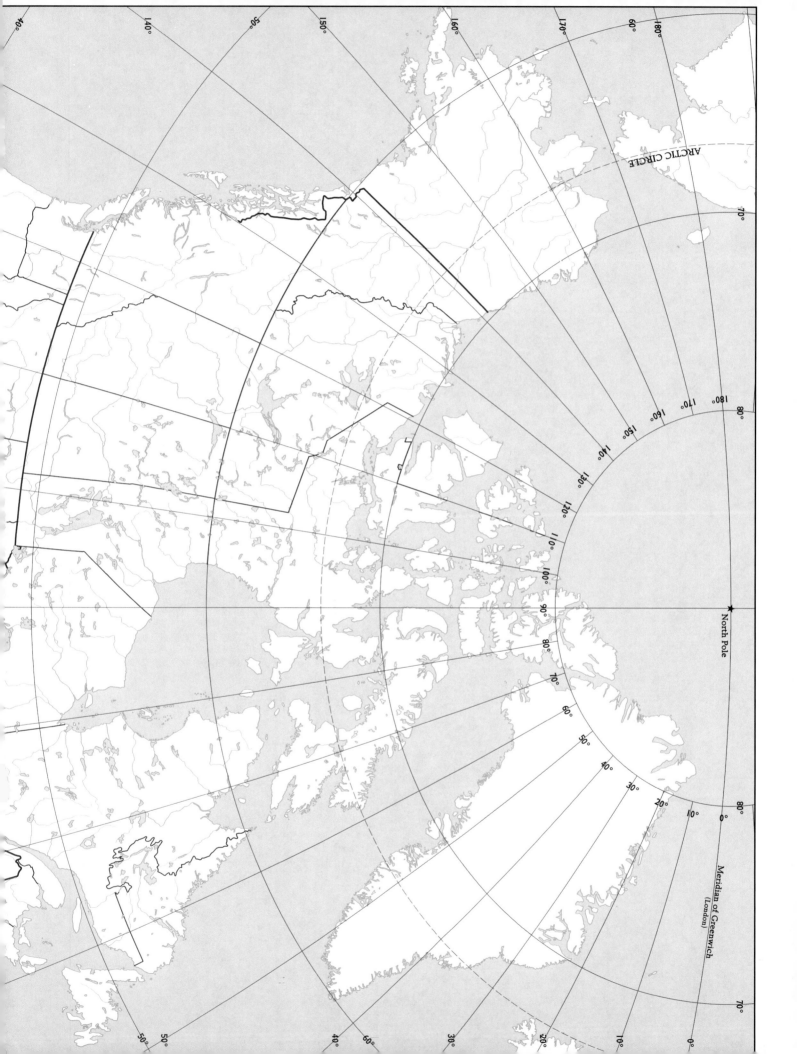

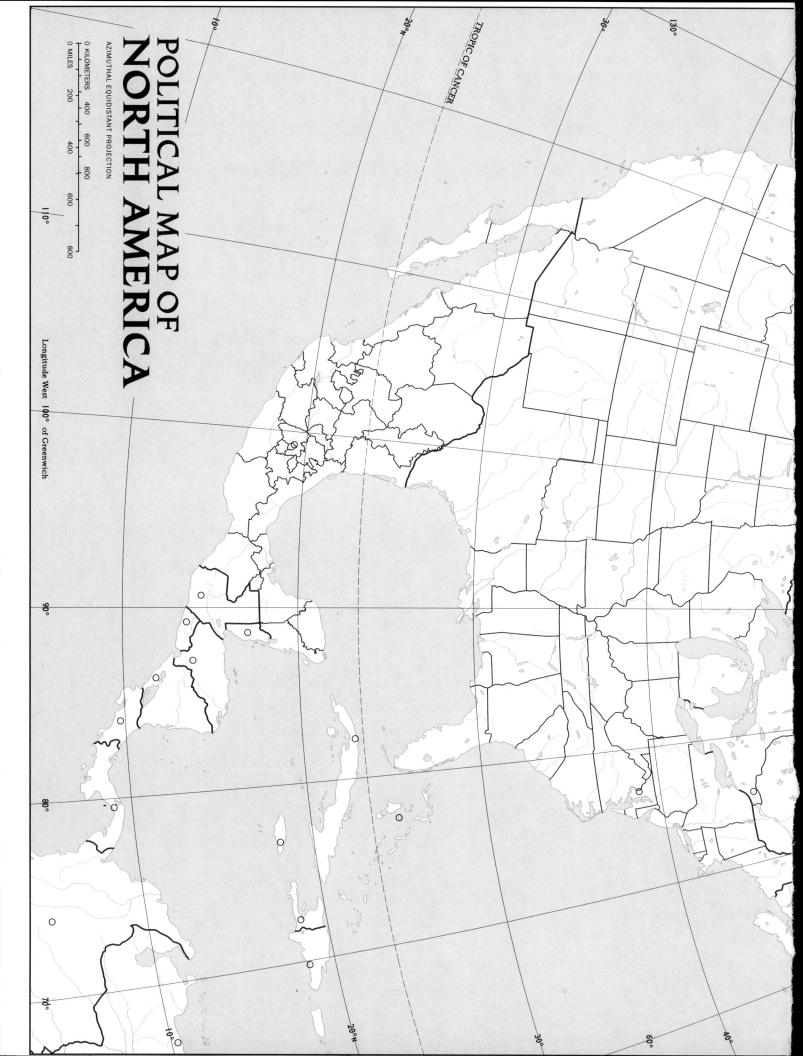

POLITICAL MAP OF
NORTH AMERICA

AZIMUTHAL EQUIDISTANT PROJECTION

0 KILOMETERS 200 400 600 800
0 MILES 400 600 800

Longitude West 100° of Greenwich

TROPIC OF CANCER

130°

30°

20°N

10°

100°

90°

80°

70°

60°

40°

20°N

10°

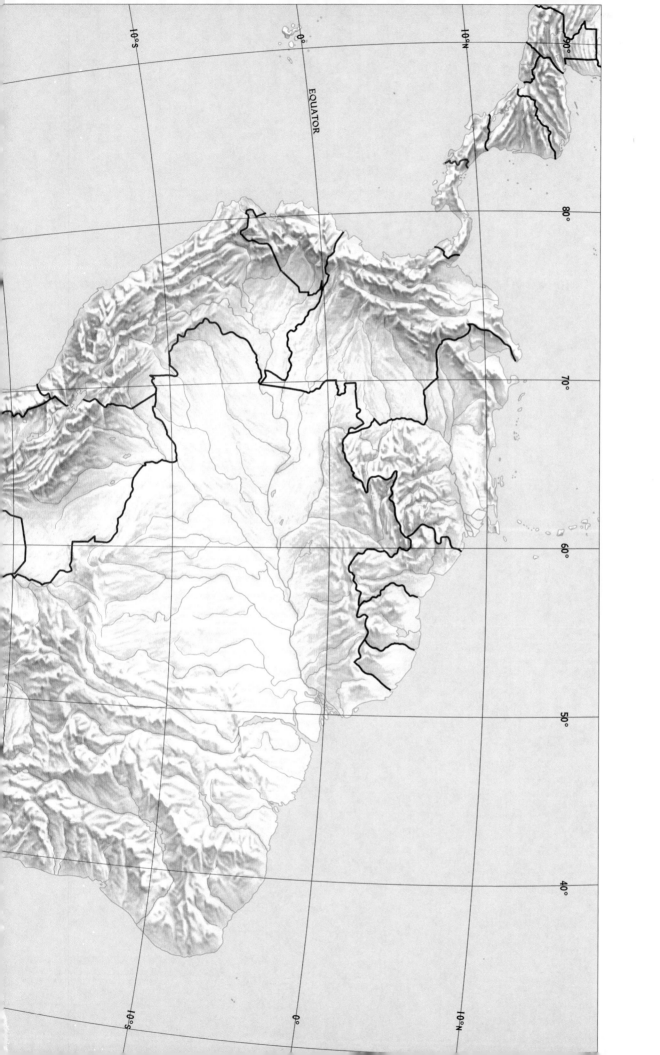

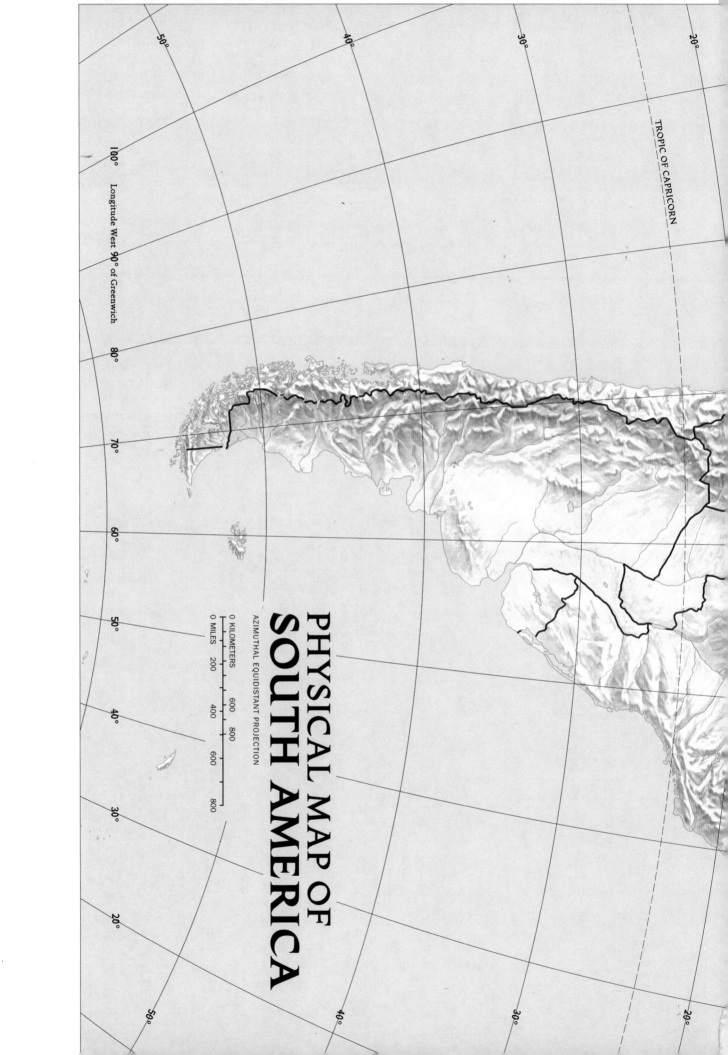

PHYSICAL MAP OF
SOUTH AMERICA

AZIMUTHAL EQUIDISTANT PROJECTION

| 0 KILOMETERS | 200 | 600 | 400 | 600 | 800 |

0 MILES   200   400   800

Longitude West 90° of Greenwich

TROPIC OF CAPRICORN

50°   40°   30°   20°

100°   80°   60°   50°   40°   30°   20°   50°

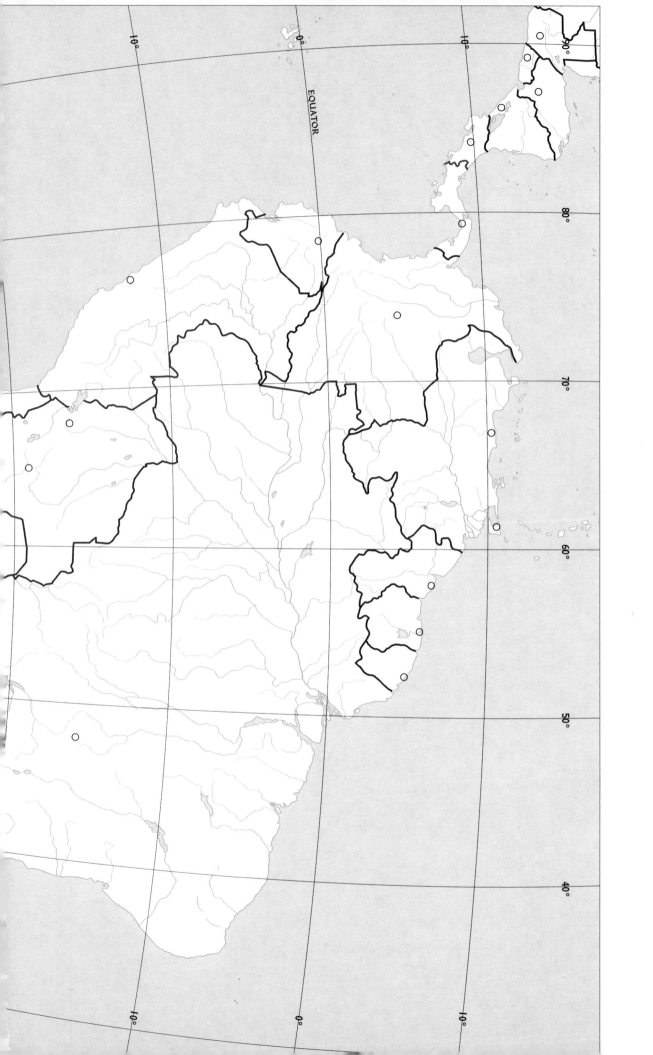

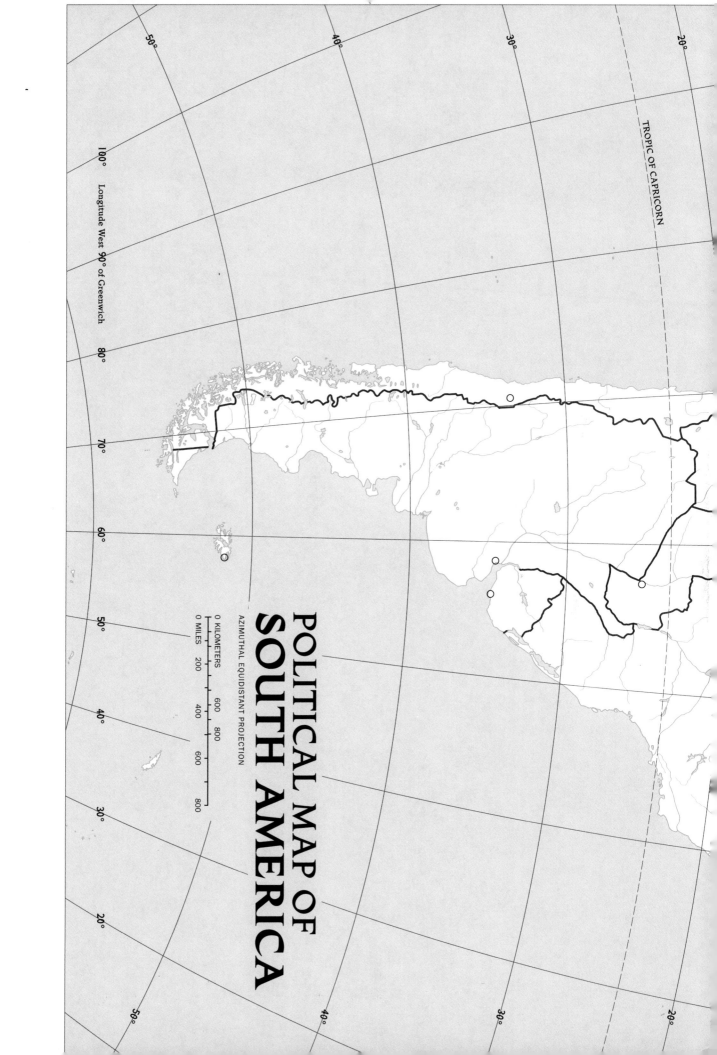

POLITICAL MAP OF
SOUTH AMERICA

AZIMUTHAL EQUIDISTANT PROJECTION

0 KILOMETERS   200   600   400   800   600   800

0 MILES

TROPIC OF CAPRICORN

Longitude West 90° of Greenwich

14_S._AMERICA_POL

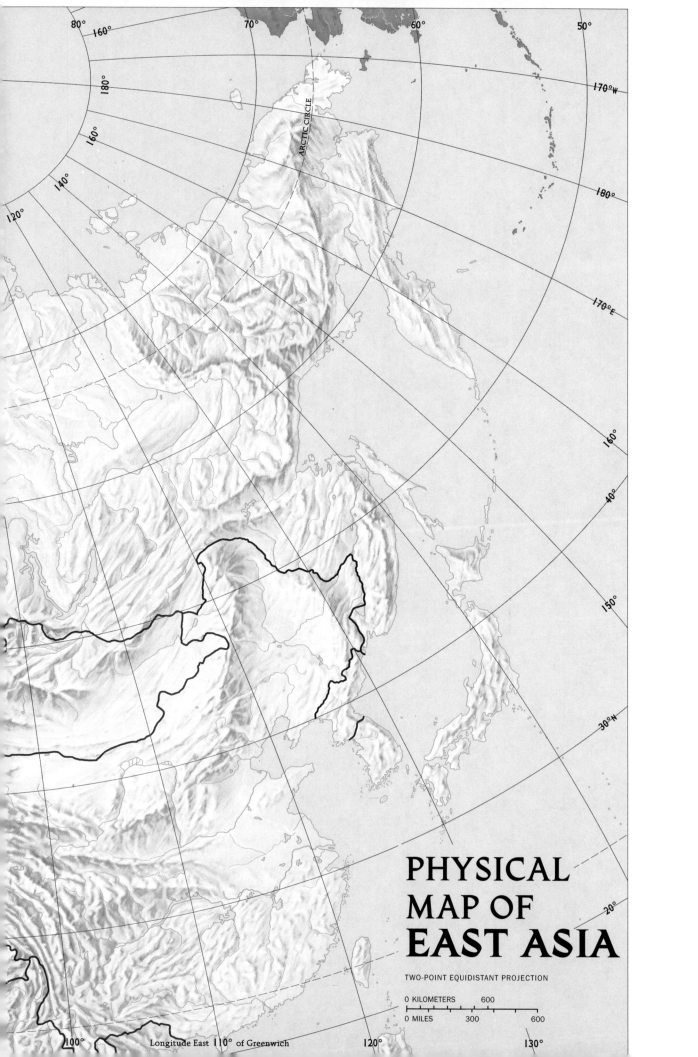

80°
160°
70°
60°
50°
170° W
180°
60°
40°
160°
ARCTIC CIRCLE
140°
120°
170° E
160°
40°
150°
30° N

PHYSICAL
MAP OF
EAST ASIA

TWO-POINT EQUIDISTANT PROJECTION

20°

| 0 KILOMETERS | 600 |
| 0 MILES | 300 | 600 |

100°
Longitude East 110° of Greenwich
120°
130°

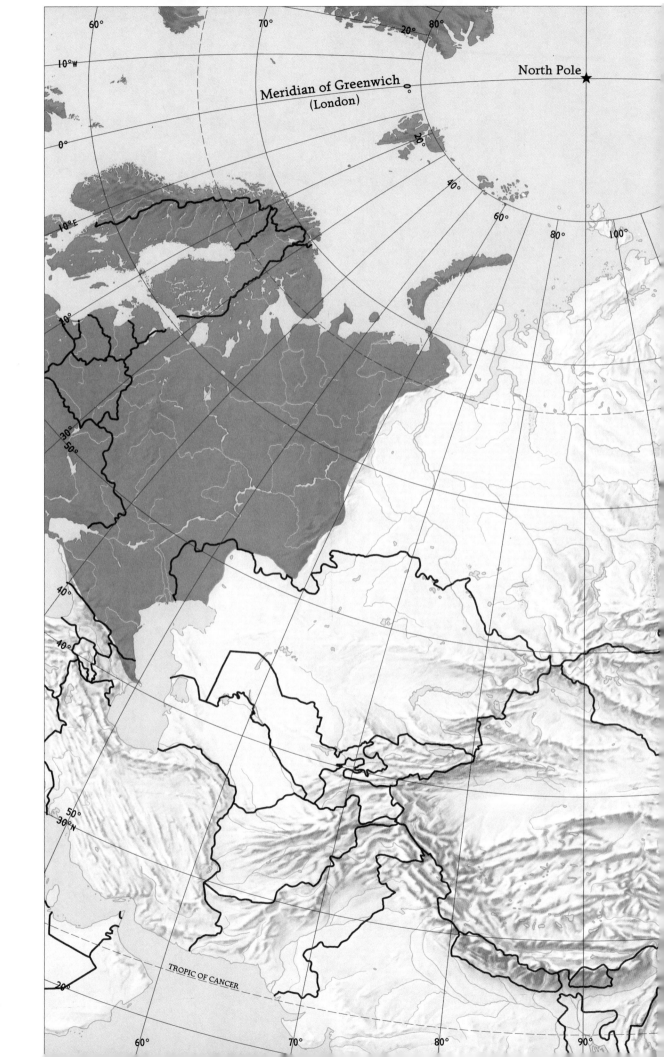

North Pole

Meridian of Greenwich
(London)

TROPIC OF CANCER

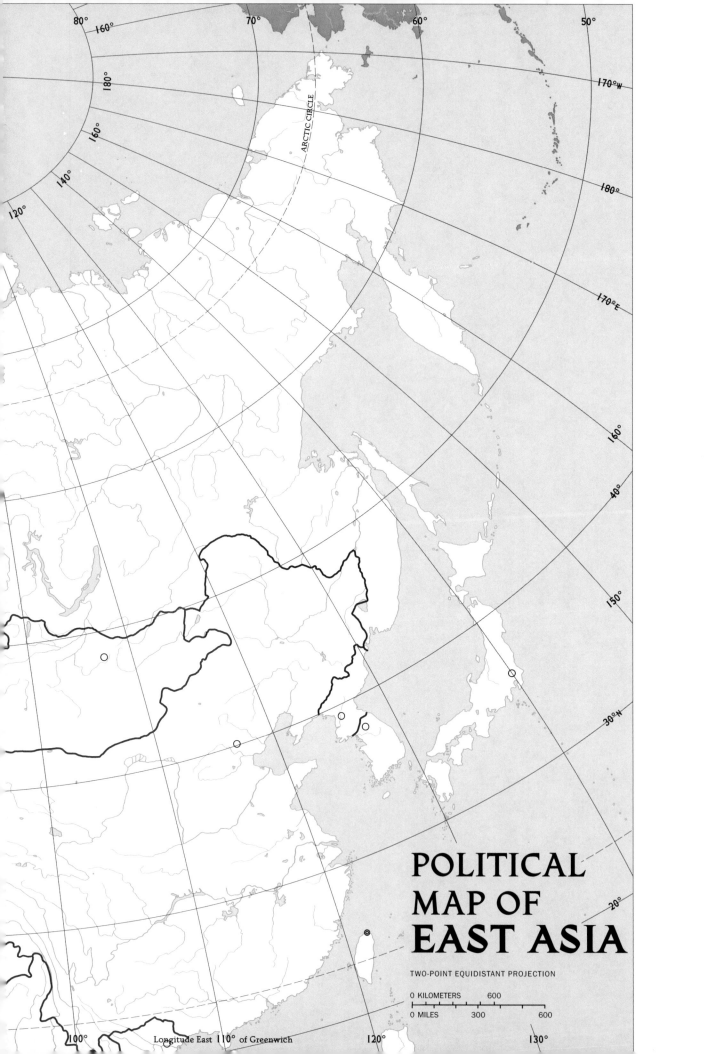

POLITICAL MAP OF EAST ASIA

TWO-POINT EQUIDISTANT PROJECTION

0 KILOMETERS		600
0 MILES	300	600

ARCTIC CIRCLE

Longitude East 110° of Greenwich

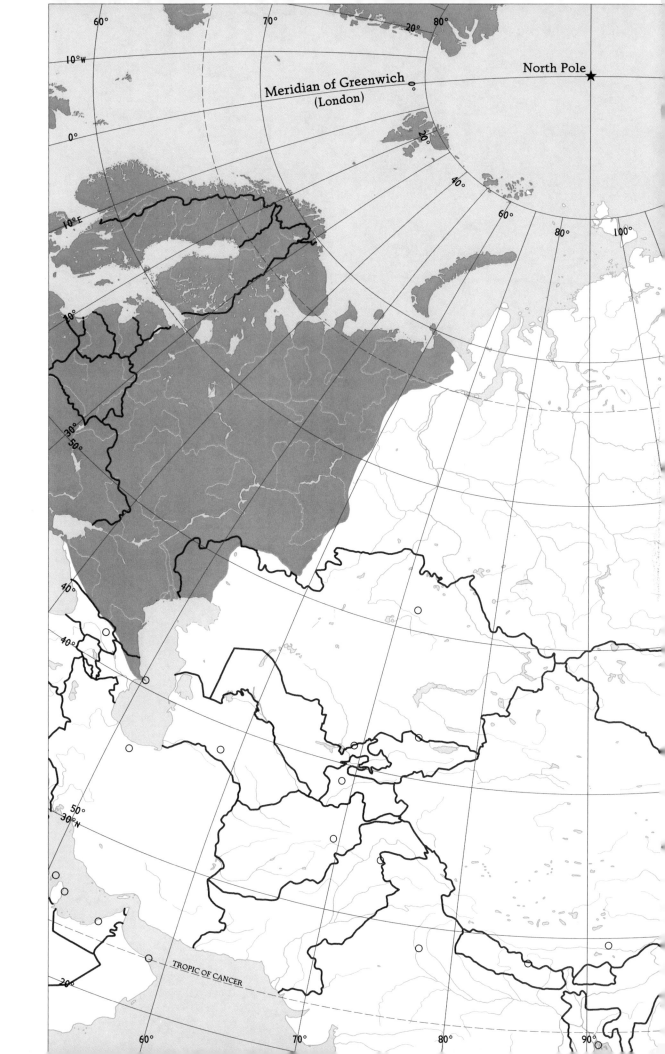

North Pole

Meridian of Greenwich
(London)

TROPIC OF CANCER

PHYSICAL MAP OF
SOUTH ASIA

LAMBERT CONFORMAL CONIC PROJECTION

0 KILOMETERS 300 400

0 MILES 200 300 400

EQUATOR

60°

66°

72°

78°

84°

90°

18°

12°

6°N

0°

6°S

12°

6°N

0°

6°S

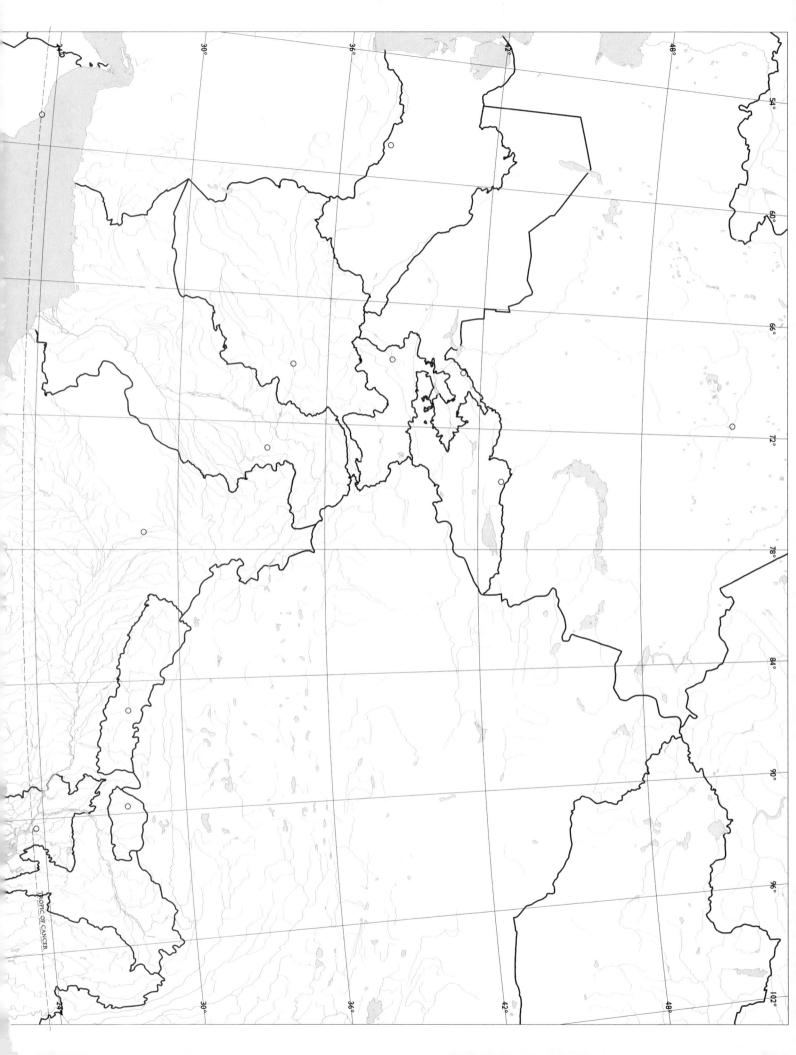

TROPIC OF CANCER

POLITICAL MAP OF
SOUTH ASIA

LAMBERT CONFORMAL CONIC PROJECTION

0 KILOMETERS 300 400

0 MILES 200 300 400

Longitude East 78° of Greenwich

EQUATOR

60°

66°

72°

84°

90°

6°S

0°

6°N

12°

18°

126°   132°   138°   144°   150°   156°

24°

TROPIC OF CANCER

18°

12°

6° N

EQUATOR   0°

6° S

126°   132°   138°   144°   150°   12°

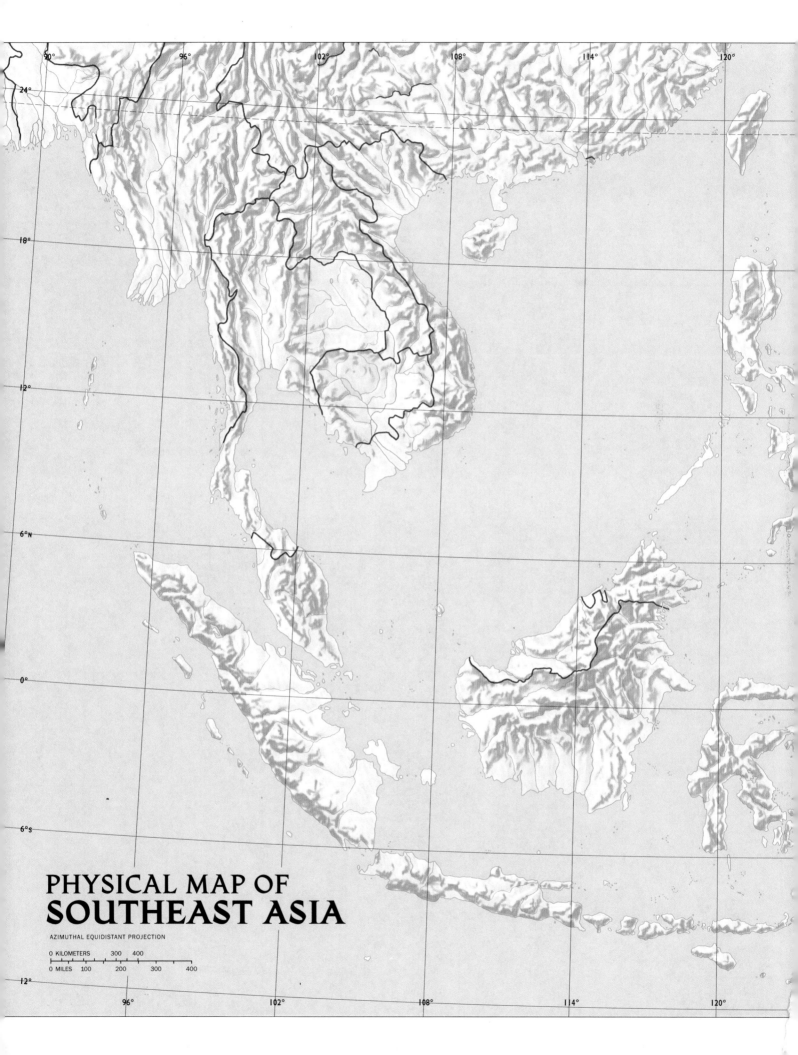

PHYSICAL MAP OF
SOUTHEAST ASIA

AZIMUTHAL EQUIDISTANT PROJECTION

0 KILOMETERS   300 400

0 MILES   100   200   300   400

126°   132°   138°   144°   150°   156°

24°

TROPIC OF CANCER

18°

12°

6°N

EQUATOR   0°

6°S

12°

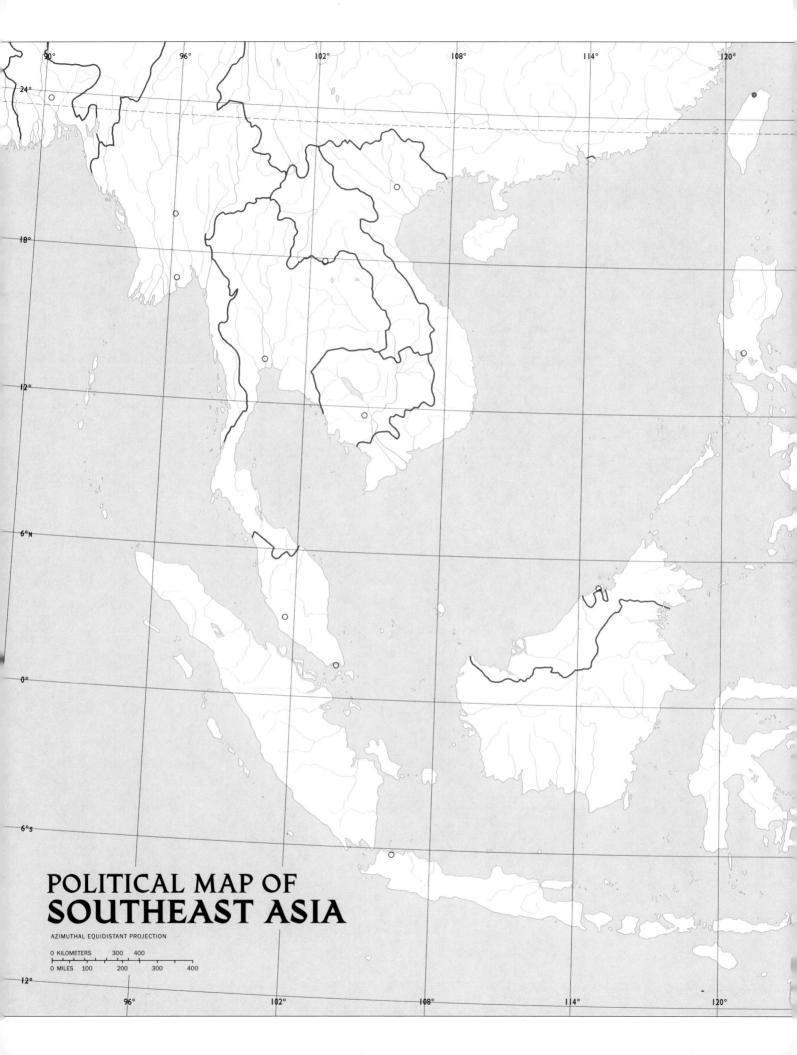

POLITICAL MAP OF
SOUTHEAST ASIA

AZIMUTHAL EQUIDISTANT PROJECTION

0 KILOMETERS   300   400

0 MILES   100   200   300   400

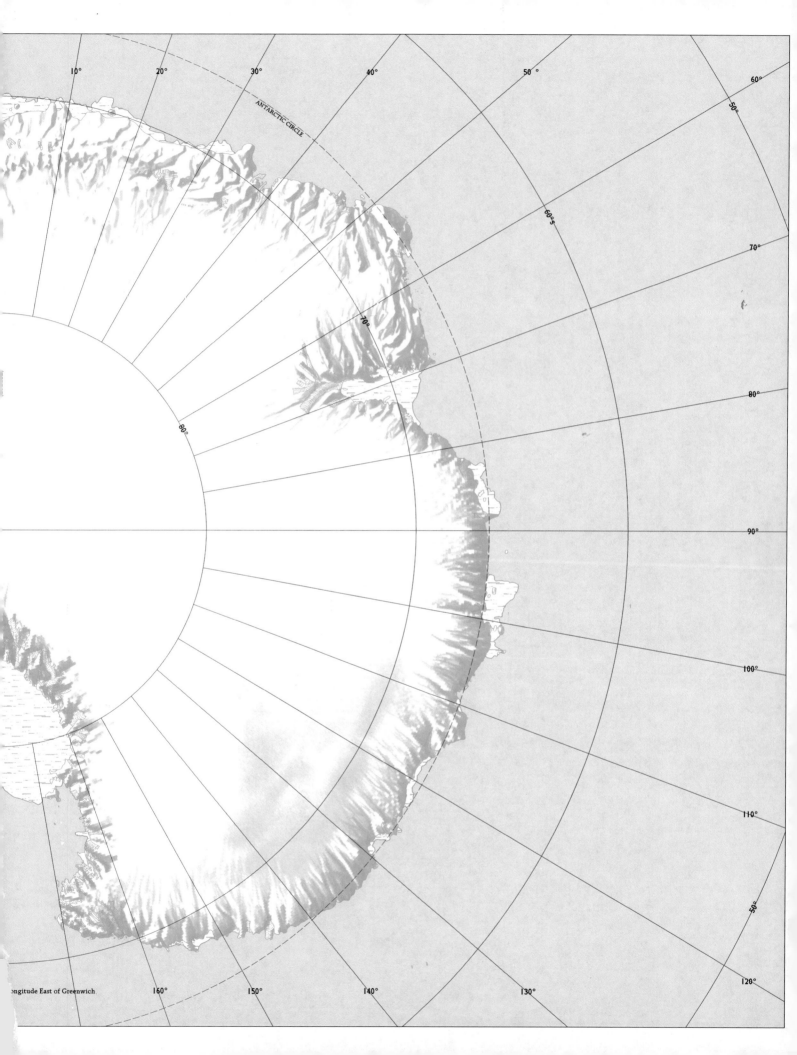

10°　20°　30°　40°　50°　60°

ANTARCTIC CIRCLE

50°

60°-S

70°

80°

70°

80°

80°

90°

100°

110°

Longitude East of Greenwich　160°　150°　140°　130°

90°

110°

120°

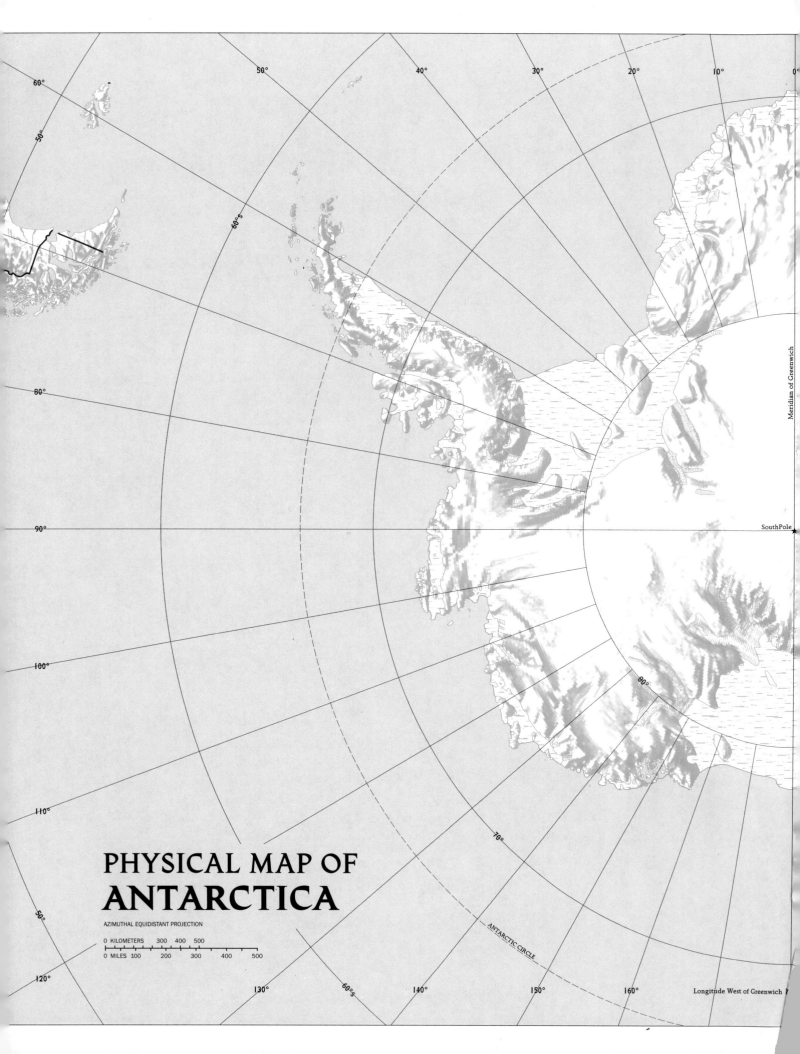

60°

50°

60°S

80°

90°

100°

110°

50°

120°

130°

60°S

140°

150°

160°

50°

40°

30°

20°

10°

0°

70°

80°

Meridian of Greenwich

SouthPole

ANTARCTIC CIRCLE

Longitude West of Greenwich

PHYSICAL MAP OF
ANTARCTICA

AZIMUTHAL EQUIDISTANT PROJECTION

0 KILOMETERS   300   400   500

0 MILES 100   200   300   400   500